ISAAC NEWTON

GREAT ASTRONOMERS

ROBERT STAWELL BALL

GIFT *Certificate*

TO:

FROM:

DATE:

Would you like to buy a copy of

ISAAC NEWTON?

PLEASE VISIT:

http://www.diamondbooks.ca

ISAAC NEWTON

GREAT ASTRONOMERS

BY

ROBERT STAWELL BALL

D.Sc, LL.D., F.R.S.

(July 1, 1840 – November 25, 1913)

LOWNDEAN PROFESSOR OF ASTRONOMY AND GEOMETRY
IN THE UNIVERSITY OF CAMBRIDGE

DIAMOND BOOKS™

www.diamondbooks.ca

TORONTO, CANADA – 2016

DIAMOND BOOKS - CANADA

DIAMOND BOOKS

Toronto, ON, CANADA

http://www.diamondbooks.ca

BIBLIOGRAPHIC INFORMATION

'ISAAC NEWTON'
Is an extract or a chapter from the book
"GREAT ASTRONOMERS" by Robert Stawell Ball.

FIRST PUBLISHED IN: 1896.

PUBLISHED IN CANADA

Published in Canada by DIAMOND BOOKS - CANADA, an imprint of DIAMOND PUBLISHERS - http://www.diamondpublishers.com

FIRST EDITION: SEPTEMBER, 2016.

PAPERBACK EDITION : ISBN: 978-1-988357-38-6

PRINTED IN CANADA

PREFACE
TO THE ORIGINAL EDITION

It has been my object in these pages to present the life of each astronomer in such detail as to enable the reader to realise in some degree the man's character and surroundings; and I have endeavoured to indicate as clearly as circumstances would permit the main features of the discoveries by which he has become known.

There are many types of astronomers—from the stargazer who merely watches the heavens, to the abstract mathematician who merely works at his desk; it has, consequently, been necessary in the case of some lives to adopt a very different treatment from that which seemed suitable for others.

ROBERT STAWELL BALL.

The Observatory, Cambridge.
October, 1895.

NOTE: The full-length of 'Preface' can be found in its original edition of the book "GREAT ASTRONOMERS".

INTRODUCTION

Of all the natural sciences there is not one which offers such sublime objects to the attention of the inquirer as does the science of astronomy. From the earliest ages the study of the stars has exercised the same fascination as it possesses at the present day. Among the most primitive peoples, the movements of the sun, the moon, and the stars commanded attention from their supposed influence on human affairs. The practical utilities of astronomy were also obvious in primeval times. Maxims of extreme antiquity show how the avocations of the husbandman are to be guided by the movements of the heavenly bodies. The positions of the stars indicated the time to plough, and the time to sow. To the mariner who was seeking a way across the trackless ocean, the heavenly bodies offered the only reliable marks by which his path could be guided. There was, accordingly, a stimulus both from intellectual curiosity and from practical necessity to follow the movements of the stars. Thus began a search for the causes of the ever-varying phenomena which the heavens display.

- ROBERT STAWELL BALL

Many of the earliest discoveries are indeed prehistoric. The great diurnal movement of the heavens, and the annual revolution of the sun, seem to have been known in times far more ancient than those to which any human monuments can be referred. The acuteness of the early observers enabled them to single out the more important of the wanderers which we now call planets. They saw that the star-like objects, Jupiter, Saturn, and Mars, with the more conspicuous Venus, constituted a class of bodies wholly distinct from the fixed stars among which their movements lay, and to which they bear such a superficial resemblance. But the penetration of the early astronomers went even further, for they recognized that Mercury also belongs to the same group, though this particular object is seen so rarely. It would seem that eclipses and other phenomena were observed at Babylon from a very remote period, while the most ancient records of celestial observations that we possess are to be found in the Chinese annals. The study of astronomy, in the sense in which we understand the word, may be said to have commenced under the reign of the Ptolemies at Alexandria. The most famous name in the science of this period is that of Hipparchus, who lived and worked at Rhodes about the year 160 B.C. It was his splendid investigations that first wrought the observed facts into a coherent branch of knowledge. He recognized the primary obligation which lies on the student of the heavens to compile as complete an inventory as possible of the objects which are there to be found. Hipparchus accordingly commenced by undertaking, on a small scale, a task exactly similar to that on which modern astronomers, with all available appliances of meridian circles, and photographic telescopes, are constantly engaged at the present day. He compiled a catalogue of the principal fixed stars, which is of special value to astronomers, as being the earliest work of its kind which has been handed down. He also studied the movements of the sun and the moon, and framed theories to account for the incessant changes which he saw in progress. He found a much more difficult problem in his attempt to interpret satisfactorily the complicated movements of the planets. With the view of constructing a theory which should give some coherent account of the subject, he made many observations of the places of these wandering stars. How great were the advances which Hipparchus accomplished may be appreciated if we reflect that, as a preliminary task to his more purely astronomical labours, he had to invent that branch of mathematical science by which alone the problems he proposed could be solved. It was for this purpose that he devised the indispensable method of calculation which we now know so well as trigonometry. Without the aid rendered by this beautiful art it would have been impossible for any n ally important advance in astronomical calculation to have been effected.

But the discovery which shows, beyond all others, that Hipparchus possessed one of the master-minds of all time was the detection of that remarkable celestial movement known as the precession of the equinoxes. The inquiry which conducted to this discovery involved a most profound investigation, especially when it is remembered that in the days of Hipparchus the means of observation of the heavenly bodies were only of the rudest description, and the available observations of earlier dates were extremely scanty. We can but look with astonishment on the genius of the man who, in spite of such difficulties, was able to detect such a phenomenon as the precession, and to exhibit its actual magnitude. I shall endeavour to explain the nature of this singular celestial movement, for it may be said to offer the first instance in the history of science in which we find that combination of accurate observation with skilful interpretation, of which, in the subsequent development of astronomy, we have so many splendid examples.

The word equinox implies the condition that the night is equal to the day. To a resident on the equator the night is no doubt equal to the day at all times in the year, but to one who lives on any other part of the earth, in either hemisphere, the night and the day are not generally equal. There is, however, one occasion in spring, and another in autumn, on which the day and the night are each twelve hours at all places on the earth. When the night and day are equal in spring, the point which the sun occupies on. the heavens is termed the vernal equinox. There is similarly another point in which the sun is situated at the time of the autumnal equinox. In any investigation of the celestial movements the positions of these two equinoxes on the heavens are of primary importance, and Hipparchus, with the instinct of genius, perceived their significance, and commenced to study them. It will be understood that we can always define the position of a point on the sky with reference to the surrounding stars. No doubt we do not see the stars near the sun when the sun is shining, but they are there nevertheless. The ingenuity of Hipparchus enabled him to determine the positions of each of the two equinoxes relatively to the stars which lie in its immediate vicinity. After examination of the celestial places of these points at different periods, he was led to the conclusion that each equinox was moving relatively to the stars, though that movement was so slow that twenty five thousand years would necessarily elapse before a complete circuit of the heavens was accomplished. Hipparchus traced out this phenomenon, and established it on an impregnable basis, so that all astronomers have ever since recognised the precession of the equinoxes) as one of the fundamental facts of astronomy. Not until nearly two thousand years after Hipparchus had made this splendid discovery was the explanation of its cause given by Newton.

From the days of Hipparchus down to the present hour the science of astronomy has steadily grown. One great observer after another has appeared from time to time, to reveal some new phenomenon with regard to the celestial bodies or their movements, while from time to time one commanding intellect after another has arisen to explain the true import of the facts of observations. The history of astronomy thus becomes inseparable from the history of the great men to whose labours its development is due. In the ensuing chapters we have endeavoured to sketch the lives and the work of the great philosophers, by whose labours the science of astronomy has been created.

We shall commence with Ptolemy, who, after the foundations of the science had been laid by Hipparchus, gave to astronomy the form in which it was taught throughout the Middle Ages. We shall next see the mighty revolution in our conceptions of the universe which are associated with the name of Copernicus. We then pass to those periods illumined by the genius of Galileo and Newton, and afterwards we shall trace the careers of other more recent discoverers, by whose industry and genius the boundaries of human knowledge have been so greatly extended. Our history will be brought down late enough to include some of the illustrious astronomers who laboured in the generation which has just passed away.

- ROBERT STAWELL BALL

TABLE OF CONTENTS

CHAPTERS

**An original Chapter from the book:*
"GREAT ASTRONOMERS", by Robert Stawell Ball.

"I do not know what I may appear to the world, but to myself I seem to have been only like a boy playing on the sea-shore, and diverting myself in now and then finding a smoother pebble or a prettier shell than ordinary, whilst the great ocean of truth lay all undiscovered before me."

- Issac Newton

CHAPTER

6

GREAT ASTRONOMERS

Isaac Newton

(December 25, 1642 – March 20, 1726/27)

Isaac Newton

(December 25, 1642 – March 20, 1726/27)

CHAPTER 6

GREAT ASTRONOMERS: ISAAC NEWTON

IT was just a year after the death of Galileo, that an infant came into the world who was christened Isaac Newton. Even the great fame of Galileo himself must be relegated to a second place in comparison with that of the philosopher who first expounded the true theory of the universe.

Isaac Newton was born on the 25th of December (old style), 1642, at Woolsthorpe, in Lincolnshire, about a half-mile from Colsterworth, and eight miles south of Grantham. His father, Mr. Isaac Newton, had died a few months after his marriage to Harriet Ayscough, the daughter of Mr. James Ayscough, of

Market Overton, in Rutlandshire. The little Isaac was at first so excessively frail and weakly that his life was despaired of. The watchful mother, however, tended her delicate child with such success that he seems to have thriven better than might have been expected from the circumstances of his infancy, and he ultimately acquired a frame strong enough to outlast the ordinary span of human life.

For three years they continued to live at Woolsthorpe, the widow's means of livelihood being supplemented by the income from another small estate at Sewstern, in a neighbouring part of Leicestershire. In 1645, Mrs. Newton took as a second husband the Rev. Barnabas Smith, and on moving to her new home, about a mile from Woolsthorpe, she entrusted little Isaac to her mother, Mrs. Ayscough. In due time we find that the boy was sent to the public school at Grantham, the name of the master being Stokes.

Woolsthorpe Manor.
Showing solar dial made by Newton when a boy.

For the purpose of being near his work, the embryo philosopher was boarded at the house of Mr. Clark, an apothecary at Grantham. We learn from Newton himself that at first he had a very low place in the class lists of the school, and was by no means one of those model school-boys who find favour in the eyes of the school-master by attention to Latin grammar. Isaac's first incentive to diligent study seems to have been derived from the circumstance that he was severely kicked by one of the boys who was above him in the class. This indignity had the effect of stimulating young Newton's activity to such an extent that he not only attained the desired object of passing over the head of the boy who had maltreated him, but continued to rise until he became the head of the school.

The play-hours of the great philosopher were devoted to pursuits very different from those of most school-boys. His chief amusement was found in making mechanical toys and various ingenious contrivances. He watched day by day with great interest the workmen engaged in constructing a windmill in the neighbourhood of the school, the result of which was that the boy made a working model of the windmill and of its machinery, which seems to have been much admired, as indicating his aptitude for mechanics. We are told that Isaac also indulged in somewhat higher flights of mechanical enterprise. He constructed a carriage, the wheels of which were to be driven by the hands of the occupant, while the first philosophical instrument he made was a clock, which was actuated by water. He also devoted much attention to the construction of paper kites, and his skill in this respect was highly appreciated by his schoolfellows. Like a true philosopher, even at this stage he experimented on the best methods of attaching the string, and on the proportions which the tail ought to have. He also made lanthorns of paper to provide himself with light as he walked to school in the dark winter mornings.

The only love affair in Newton's life appears to have commenced while he was still of tender years. The incidents are thus described in Brewster's "Life of Newton," a work to which I am much indebted in this chapter.

"In the house where he lodged there were some female inmates, in whose company he appears to have taken much pleasure. One of these, a Miss Storey, sister to Dr.

Storey, a physician at Buckminster, near Colsterworth, was two or three years younger than Newton, and to great personal attractions she seems to have added more than the usual allotment of female talent. The society of this young lady and her companions was always preferred to that of his own schoolfellows, and it was one of his most agreeable occupations to construct for them little tables and cupboards, and other utensils for holding their dolls and their trinkets. He had lived nearly six years in the same house with Miss Storey, and there is reason to believe that their youthful friendship gradually rose to a higher passion; but the smallness of her portion, and the inadequacy of his own fortune, appear to have prevented the consummation of their happiness. Miss Storey was afterwards twice married, and under the name of Mrs. Vincent, Dr. Stukeley visited her at Grantham in 1727, at the age of eighty-two, and obtained from her many particulars respecting the early history of our author. Newton's esteem for her continued unabated during his life. He regularly visited her when he went to Lincolnshire, and never failed to relieve her from little pecuniary difficulties which seem to have beset her family."

The schoolboy at Grantham was only fourteen years of age when his mother became a widow for the second time. She then returned to the old family home at Woolsthorpe, bringing with her the three children of her second marriage. Her means appear to have been somewhat scanty, and it was consequently thought necessary to recall Isaac from the school. His recently-born industry had been such that he had already made good progress in his studies, and his mother hoped that he would now lay aside his books, and those silent meditations to which, even at this early age, he had become addicted. It was expected that, instead of such pursuits, which were deemed quite useless, the boy would enter busily into the duties of the farm and the details of a country life. But before long it became manifest that the study of nature and the pursuit of knowledge had such a fascination for the youth that he could give little attention to aught else. It was plain that he would make but an indifferent farmer. He greatly preferred experimenting on his water-wheels to looking after labourers, while he found that working at mathematics behind a hedge was much more interesting than chaffering about the price of bullocks in the marketplace. Fortunately for humanity his mother, like a wise woman, determined to let her boy's genius have the scope which it required. He was accordingly sent back to Grantham school, with the object of being trained in the knowledge which would fit him for entering the university of Cambridge.

It was the 5th of June, 1660, when Isaac Newton, a youth of eighteen, was enrolled as an undergraduate of Trinity College, Cambridge. Little did those who sent him

Trinity College, Cambridge.

Showing Newton's rooms; on the leads of the gateway he placed his telescope.

- ROBERT STAWELL BALL

there dream that this boy was destined to be the most illustrious student who ever entered the portals of that great seat of learning. Little could the youth himself have foreseen that the rooms near the gateway which he occupied would acquire a celebrity from the fact that he dwelt in them, or that the ante-chapel of his college was in good time to be adorned by that noble statue, which is regarded as one of the chief art treasures of Cambridge University, both on account of its intrinsic beauty and the fact that it commemorates the fame of her most distinguished alumnus, Isaac Newton, the immortal astronomer. Indeed, his advent at the University seemed to have been by no means auspicious or brilliant. His birth was, as we have seen, comparatively obscure, and though he had already given indication of his capacity for reflecting on philosophical matters, yet he seems to have been but ill-equipped with the routine knowledge which youths are generally expected to take with them to the Universities. From the outset of his college career, Newton's attention seems to have been mainly directed to mathematics. Here he began to give evidence of that marvellous insight into the deep secrets of nature which more than a century later led so dispassionate a judge as Laplace to pronounce Newton's immortal work as pre-eminent above all the productions of the human intellect. But though Newton was one of the very greatest mathematicians that ever lived, he was never a mathematician for the mere sake of mathematics. He employed his mathematics as an instrument for discovering the laws of nature. His industry and genius soon brought him under the notice of the University authorities. It is stated in the University records that he obtained a Scholarship in 1664. Two years later we find that Newton, as well as many residents in the University, had to leave Cambridge temporarily on account of the breaking out of the plague. The philosopher retired for a season to his old home at Woolsthorpe, and there he remained until he was appointed a Fellow of Trinity College, Cambridge, in 1667. From this time onwards, Newton's reputation as a mathematician and as a natural philosopher steadily advanced, so that in 1669, while still but twenty-seven years of age, he was appointed to the distinguished position of Lucasian Professor of Mathematics at Cambridge. Here he found the opportunity to continue and develop that marvellous career of discovery which formed his life's work.

The earliest of Newton's great achievements in natural philosophy was his detection of the composite character of light. That a beam of ordinary sunlight is, in fact, a mixture of a very great number of different-coloured lights, is a doctrine now familiar to everyone who has the slightest education in physical science. We must, however, remember that this discovery was really a tremendous advance in knowledge at the time when Newton announced it.

We here give the little diagram originally drawn by Newton, to explain the experiment by which he first learned the composition of light.

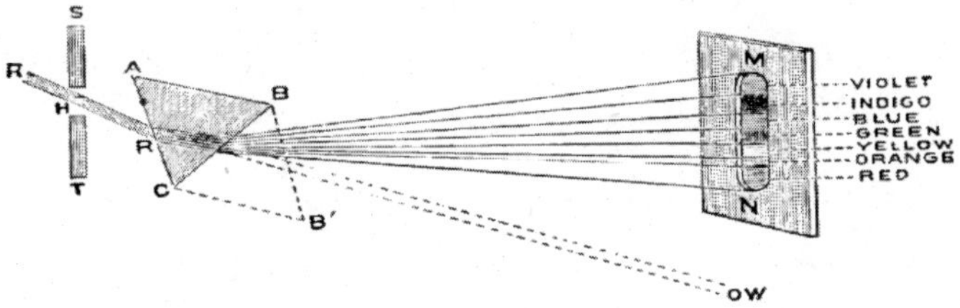

A sunbeam is admitted into a darkened room through an opening, H, in a shutter. This beam when not interfered with will travel in a straight line to the screen, and there reproduce a bright spot of the same shape as the hole in the shutter. If, however, a prism of glass, A B C, be introduced so that the beam traverse it, then it will be seen at once that the light is deflected from its original track. There is, however, a further and most important change which takes place. The spot of light is not alone removed to another part of the screen, but it becomes spread out into a long-band beautifully coloured, and exhibiting the hues of the rainbow. At the top are the violet rays, and then in descending order we have the indigo, blue, green, yellow, orange, and red.

The circumstance in this phenomenon which appears to have particularly arrested Newton's attention, was the elongation which the luminous spot underwent in consequence of its passage through the prism. When the prism was absent the spot was nearly circular, but when the prism was introduced the spot was about five times as long as it was broad. To ascertain the explanation of this was the first problem to be solved. It seemed natural to suppose that it might be due to the thickness of the glass in the prism which the light traversed, or to the angle of incidence at which the light fell upon the prism. He found, however, upon careful trial, that the phenomenon could not be thus accounted for. It was not until after much patient labour that the true explanation dawned upon him. He discovered that though the beam of white light looks so pure and so simple, yet in reality it is composed of differently coloured lights blended together. These are, of course, indistinguishable in the compound beam, but they are separated or disentangled, so to speak, by the action of the prism. The rays at the blue end of the spectrum are more powerfully deflected by the action of the glass than are the rays at the red end. Thus, the rays variously coloured red, orange, yellow, green, blue, indigo, violet, are each conducted to a different part of the screen. In this way the prism has the effect of exhibiting the constitution of the composite beam of light. To us this now seems quite obvious, but Newton did not adopt it hastily. With characteristic caution he verified the explanation by many different experiments, all of which confirmed his discovery. One of these may be mentioned.

He made a hole in the screen at that part on which the violet rays fell. Thus a violet ray was al-lowed to pass through, all the rest of the light being intercepted, and on this beam so isolated he was able to try further experiments. For instance, when he interposed another prism in its path, he found, as he expected, that it was again deflected, and he measured the amount of the deflection. Again he tried the same experiment with one of the red rays from the opposite end of the coloured band. He allowed it to pass through the same aperture in the screen, and he tested the amount by which the second prism was capable of producing deflection. He thus found, as he had expected to find, that the second prism was more efficacious in bending the violet rays than in bending the red rays. Thus he confirmed the fact that the various hues of the rainbow were each bent by a prism to a different extent, violet being acted upon the most, and red the least.

Not only did Newton decompose a white beam into its constituent colours, but conversely by interposing a second prism with its angle turned upwards, he reunited the different col-ours, and thus reproduced the original beam of white light. In several other ways also he illustrated his famous proposition, which then seemed so startling, that white light was the result of a mixture of all hues of the rainbow. By combining painters' colours in the proper proportion he did not indeed succeed in producing a mixture which would ordinarily be called white, but he obtained a grey pigment. Some of this he put on the floor of his room for comparison with a piece of white paper. He allowed a beam of bright sunlight to fall upon the paper and the mixed colours side by side, and a friend whom he called in for his opinion pronounced that under these circumstances the mixed colours looked the whiter of the two.

- ROBERT STAWELL BALL

By repeated demonstrations Newton thus established his great discovery of the composite character of light. He at once perceived that his researches had an important bearing upon the principles involved in the construction of a telescope. Those who employed the telescope for looking at the stars, had been long aware of the imperfections which prevented all the various rays from being conducted to the same focus. But this imperfection had hitherto been erroneously accounted for. It had been supposed that the reason why success had not been attained in the construction of a refracting telescope was due to the fact that the object glass, made as it then was of a single piece, had not been properly shaped. Mathematicians had abundantly demonstrated that a single lens, if properly figured, must conduct all rays of light to the same focus, provided all rays experienced equal refraction in passing through the glass. Until Newton's discovery of the composition of white light, it had been taken for granted that the several rays in a white beam were equally refrangible. No doubt if this had been the case, a perfect telescope could have been produced by properly shaping the object glass. But when Newton had demonstrated that light was by no means so simple as had been supposed, it became obvious that a satisfactory refracting telescope was an impossibility when only a single object lens was employed, however carefully that lens might have been wrought. Such an objective might, no doubt, be made to conduct any one group of rays of a particular shade to the same focus, but the rays of other colours in the beam of white light must necessarily travel somewhat astray. In this way Newton accounted for a great part of the difficulties which had hitherto beset the attempts to construct a perfect refracting telescope.

We now know how these difficulties can be, to a great extent, overcome, by employing for the objective a composite lens made of two pieces of glass possessing different qualities. To these achromatic object glasses, as they are called, the great development of astronomical knowledge, since Newton's time, is due. But it must be remarked that, although the theoretical possibility of constructing an achromatic lens was investigated by Newton, he certainly came to the conclusion that the difficulty could not be removed by employing a composite objective with two different kinds of glass. In this his marvellous sagacity in the interpretation of nature seems for once to have deserted him. We can, however, hardly regret that Newton failed to discover the achromatic objective, when we observe that it was in consequence of his deeming an achromatic objective to be impossible that he was led to the invention of the reflecting telescope.

Finding, as he believed, that the defects of the telescope could not be remedied by any application of the principle of refraction, he was led to look in quite a different direction for the improvement of the tool on which the advancement of astronomy depended. The *refraction* of light depended, as he had found, upon the colour of the light. The laws of *reflection* were, however, quite independent of the colour. Whether rays be red or green, blue or yellow, they are all reflected in precisely the same manner from a mirror. Accordingly, Newton perceived that if he could construct a telescope the action of which depended upon reflection, instead of upon refraction, the difficulty which had hitherto proved an insuperable obstacle to the improvement of the instrument would be evaded. For this purpose Newton fashioned a concave mirror from a mixture of copper and tin, a combination which gives a surface with almost the lustre of silver.

- ROBERT STAWELL BALL

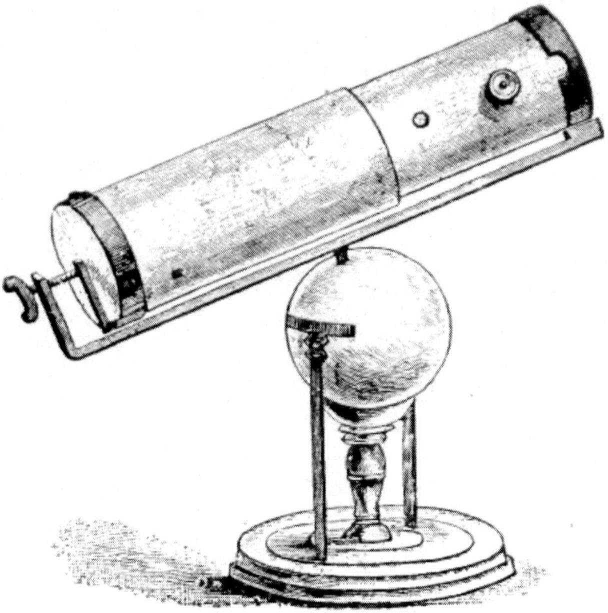

Sir Isaac Newton's little Reflector.

When the light of a star fell upon the surface, an image of the star was produced in the focus of this mirror, and then this image was examined by a magnifying eye-piece. Such is the principle of the famous reflecting telescope which bears the name of Newton. The little reflector which he constructed, represented in the adjoining figure, is still preserved as one of the treasures of the Royal Society. The telescope tube had the very modest dimension of one inch in diameter. It was, however, the precursor of a whole series of magnificent instruments, each outstripping the other in magnitude, until at last the culminating point was attained in 1845, by the construction of Lord Rosse's mammoth reflector of six feet in aperture.

Newton's discovery of the composition of light led to an embittered controversy, which caused no little worry to the great philosopher. Some of those who attacked him enjoyed considerable and, it must be admitted, even well-merited repute in the ranks of science. They alleged, however, that the elongation of the coloured band which Newton had noticed was due to this, to that, or to the other—to anything, in fact, rather than to the true cause which Newton assigned. With characteristic patience and love of truth, Newton steadily replied to each such attack. He showed most completely how utterly his adversaries had misunderstood the subject, and how slight in-deed was their acquaintance with the natural phenomenon in question. In reply to each point raised, he was ever able to cite fresh experiments and adduce fresh illustrations, until at last his opponents retired worsted from the combat.

It has been often a matter for surprise that Newton, throughout his whole career, should have taken so much trouble to expose the errors of those who attacked his views. He used even to do this when it plainly appeared that his adversaries did not understand the subject they were discussing. A philosopher might have said, "I know I am right, and whether others think I am right or not may be a matter of concern to them, but it is certainly not a matter about which I need trouble. If after having been told the truth they elect to remain in error, so much the worse for them; my time can be better employed than in seeking to put such people right. "This, however, was not Newton's method. He spent much valuable time in overthrowing objections which were often of a very futile description. Indeed, he suffered a great deal of annoyance from the persistency, and in some cases one might almost say from the rancour, of the attacks which were made upon him. Unfortunately for himself, he did not possess that capacity for sublime indifference to what men may say, which is often the happy possession of intellects greatly inferior to his. The subject of optics still continuing to engross Newton's attention, he followed up his researches into the structure of the sunbeam by many other valuable investigations in connection with light. Everyone has noticed the beautiful colours manifested in a soap-bubble. Here was a subject which not unnaturally attracted the attention of one who had expounded the colours of the spectrum with such success. He perceived that similar hues were produced by other thin plates of transparent material besides soap-bubbles, and his ingenuity was sufficient to devise a method by which the thicknesses of the different films could be measured. We can hardly, indeed, say that a like success attended his interpretation of these phenomena to that which had been so conspicuous in his explanation of the spectrum. It implies no disparagement to the sublime genius of Newton to admit that the doctrines he put forth as to the causes of the colours in the soap-bubbles can be no longer accepted. We must remember that Newton was a pioneer in accounting for the physical properties of light. The facts that he established are indeed unquestionable, but the explanations which he was led to offer of some of them are seen to be untenable in the fuller light of our present knowledge.

Had Newton done nothing beyond making his wonderful discoveries in light, his fame would have gone down to posterity as one of the greatest of Nature's interpreters. But it was reserved for him to accomplish other discoveries, which have pushed even his analysis of the sunbeam into the background; it is he who has expounded the system of the universe by the discovery of the law of universal gravitation.

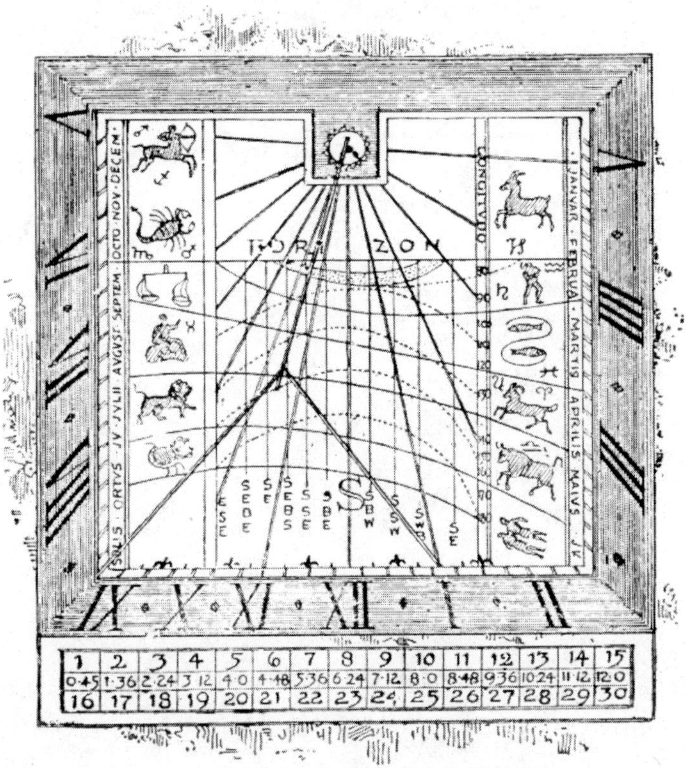

Sir Isaac Newton's Sun-dial.

- ROBERT STAWELL BALL

The age had indeed become ripe for the advent of the genius of Newton. Kepler had discovered with marvellous penetration the laws which govern the movements of the planets around the sun, and in various directions it had been more or less vaguely felt that the explanation of Kepler's laws, as well as of many other phenomena, must be sought for in connection with the attractive power of matter. But the mathematical analysis which alone could deal with this subject was wanting; it had to be created by Newton.

At Woolsthorpe, in the year 1666, Newton's attention appears to have been concentrated upon the subject of gravitation. Whatever may be the extent to which we accept the more or less mythical story as to how the fall of an apple first directed the attention of the philosopher to the fact that gravitation must extend through space, it seems, at all events, certain that this is an excellent illustration of the line of reasoning which he followed. He argued in this way. The earth attracts the apple ; it would do so, no matter how high might be the tree from which that apple fell. It would then seem to follow that this power which resides in the earth by which it can draw all external bodies towards it, extends far beyond the altitude of the loftiest tree. Indeed, we seem to find no limit to it. At the greatest elevation that has ever been attained, the attractive power of the earth is still exerted, and though we cannot by any actual experiment reach an altitude more than a few miles above the earth, yet it is certain that gravitation would extend to elevations far greater. It is plain, thought Newton, that an apple let fall from a point a hundred miles above this earth's surface, would be drawn down by the attraction, and would continually gather fresh velocity until it reached the ground. From a hundred miles it was natural to think of what would happen at a thousand miles, or at hundreds of thousands of miles. No doubt the intensity of the attraction becomes weaker with every increase in the altitude, but that action would still exist to some extent, however lofty might be the elevation which had been attained.

It then occurred to Newton, that though the moon is at a distance of two hundred and forty thousand miles from the earth, yet the attractive power of the earth must extend to the moon. He was particularly led to think of the moon in this connection, not only because the moon is so much closer to the earth than are any of the other celestial bodies, but also because the moon is an appendage to the earth, always revolving around it. The moon is certainly attracted to the earth, and yet the moon does not fall down ; how is this to be accounted for? The explanation was to be found in the character of the moon's present motion. If the moon were left for a moment at rest, there can be no doubt that the attraction to the earth would begin to draw the lunar globe in towards our globe.

In the course of a few days our satellite would come down on the earth with a most fearful crash. This catastrophe is averted by the circumstance that the moon has a movement of revolution around the earth. Newton was able to calculate from the known laws of mechanics, which he had himself been mainly instrumental in discovering, what the attractive power of the earth must be, so that the moon shall move precisely as we find it to move. It then appeared that the very power which makes an apple fall at the earth's surface, is the power which guides the moon in its orbit. Once this step had been taken, the whole scheme of the universe might almost be said to have become unrolled before the eye of the philosopher. It was natural to suppose that just as the moon was guided and controlled by the attraction of the earth, so the earth itself, in the course of its great annual progress, should guided and controlled by the supreme attractive power of the sun.

- ROBERT STAWELL BALL

Sir Isaac Newton's Telescope.

If this were so with regard to the earth, then it would be impossible to doubt that in the same way the movements of the planets could be explained to be consequences of solar attraction.

It was at this point that the great laws of Kepler became especially significant. Kepler had shown how each of the planets revolves in an ellipse around the sun, which is situated on one of the foci. This discovery had been arrived at from the interpretation of observations. Kepler had himself assigned no reason why the orbit of a planet should be an ellipse rather than any other of the infinite number of closed curves which might be traced around the sun. Kepler had also shown, and here again he was merely deducing the results from observation, that when the movements of two planets were compared together, the squares of the periodic times in which each planet revolved were proportional to the cubes of their mean distances from the sun. This also Kepler merely knew to be true as a fact, he gave no demonstration of the reason why nature should have adopted this particular relation between the distance and the periodic time rather than any other. Then, too, there was the law by which Kepler, with unparalleled ingenuity, explained the way in which the velocity of a planet varies at the different points of its track, when he showed how the line drawn from the sun to the planet described equal areas around the sun in equal times. These were the materials with which Newton set to work. He proposed to infer from these the actual laws regulating the force by which the sun guides the planets. Here it was that his sublime mathematical genius came into play. Step by step Newton advanced until he had completely accounted for all the phenomena.

In the first place, he showed that as the planet describes equal areas in equal times about the sun, the attractive force which the sun exerts upon it must necessarily be directed in a straight line towards the sun itself. He also demonstrated the converse truth, that whatever be the nature of the force which emanated from a sun, yet so long as that force was directed through the sun's centre, any body which revolved around it must describe equal areas in equal times, and this it must do, whatever be the actual character of the law according to which the intensity of the force varies at different parts of the planet's journey. Thus the first advance was taken in the exposition of the scheme of the universe.

The next step was to determine the law according to which the force thus proved to reside in the sun varied with the distance of the planet. Newton presently showed by a most superb effort of mathematical reasoning, that if the orbit of a planet were an ellipse and if the sun were at one of the foci of that ellipse, the intensity of the attractive force must vary inversely as the square of the planet's distance. If the law had any other expression than the inverse square of the distance, then the orbit which the planet must follow would not be an ellipse ; or if an ellipse, it would, at all events, not have the sun in the focus. Hence he was able to show from Kepler's laws alone that the force which guided the planets was an attractive power emanating from the sun, and that the intensity of this attractive power varied with the inverse square of the distance between the two bodies.

These circumstances being known, it was then easy to show that the last of Kepler's three laws must necessarily follow. If a number of planets were revolving around the sun, then supposing the materials of all these bodies were equally affected by gravitation, it can be demonstrated that the square of the periodic time in which each planet completes its orbit is proportional to the cube of the greatest diameter in that orbit.

These superb discoveries were, however, but the starting-point from which Newton entered on a series of researches, which disclosed many of the profoundest secrets in the scheme of celestial mechanics. His natural insight showed that not only large masses like the sun and the earth, and the moon, attract each other, but that every particle in the universe must attract every other particle with a force which varies inversely as the square of the distance between them. If, for example, the two particles were placed twice as far apart, then the intensity of the force which sought to bring them together would be

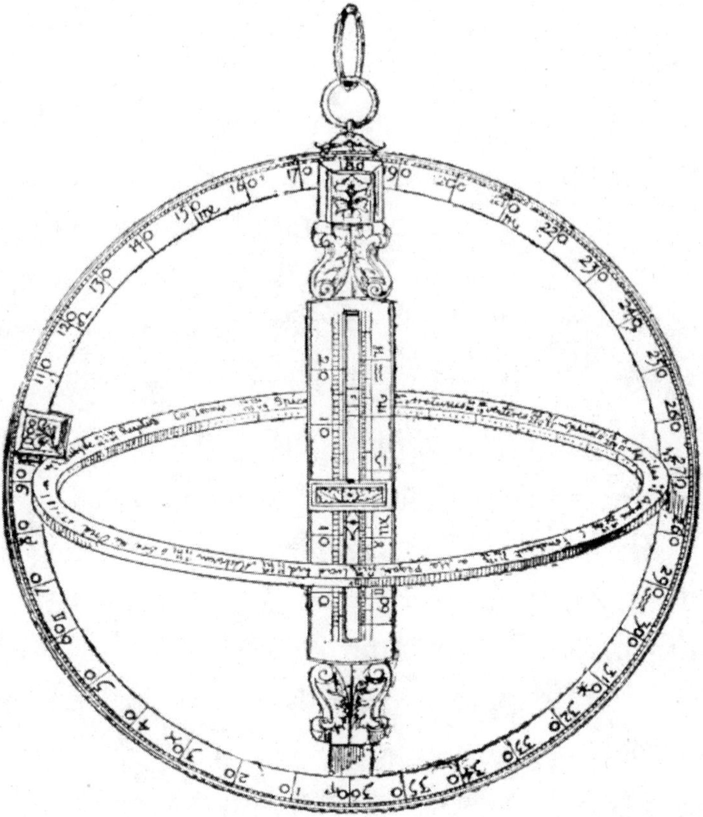

Sir Isaac Newton's Astrolabe.

reduced to one-fourth. If two particles, originally ten miles asunder, attracted each other with a certain force, then, when the distance was reduced to one mile, the intensity of the attraction between the two particles would be increased one hundred-fold. This fertile principle extends throughout the whole of nature. In some cases, however, the calculation of its effect upon the actual problems of nature would be hardly possible, were it not for another discovery which Newton's genius enabled him to accomplish.

In the case of two globes like the earth and the moon, we must remember that we are dealing not with particles, but with two mighty masses of matter, each composed of innumerable myriads of particles. Every particle in the earth does attract every particle in the moon with a force which varies inversely as the square of their distance. The calculation of such attractions is rendered feasible by the following principle. Assuming that the earth consists of materials symmetrically arranged in shells of varying densities, we may then, in calculating its attraction, regard the whole mass of the globe as concentrated at its centre. Similarly we may regard the moon as concentrated at the centre of its mass. In this way the earth and the moon can both be regarded as particles in point of size, each particle having, however, the entire mass of the corresponding globe. The attraction of one particle for another is a much more simple matter to investigate than the attraction of the myriad different points of the earth upon the myriad different points of the moon.

Many great discoveries now crowded in upon Newton. He first of all gave the explanation of the tides that ebb and flow around our shores. Even in the earliest times the tides had been shown to be related to the moon. It was noticed that the tides were specially high during full moon or during new moon, and this circumstance obviously pointed to the existence of some connection between the moon and these movements of the water, though as to what that connection was no one had any accurate conception until

Newton announced the law of gravitation. Newton then made it plain that the rise and fall of the water was simply a consequence of the attractive power which the moon exerted upon the oceans lying upon our globe. He showed also that to a certain extent the sun produces tides, and he was able to explain how it was that when the sun and the moon both conspire, the joint result was to produce especially high tides, which we call “spring tides”; whereas if the solar tide was low, while the lunar tide was high, then we had the phenomenon of “neap” tides.

But perhaps the most signal of Newton's applications of the law of gravitation was connected with certain irregularities in the movements of the moon. In its orbit round the earth our satellite is, of course, mainly guided by the great attraction of our globe. If there were no other body in the universe, then the centre of the moon must necessarily perform an ellipse, and the centre of the earth would lie in the focus of that ellipse. Nature, however, does not allow the movements to possess the simplicity which this arrangement would imply, for the sun is present as a source of disturbance. The sun attracts the moon, and the sun attracts the earth, but in different degrees, and the consequence is that the moon's movement with regard to the earth is seriously affected by the influence of the sun. It is not allowed to move exactly in an ellipse, nor is the earth exactly in the focus. How great was Newton's achievement in the solution of this problem will be appreciated if we realise that he not only had to determine from the law of gravitation the nature of the disturbance of the moon, but he had actually to construct the mathematical tools by which alone such calculations could be effected.

The resources of Newton's genius seemed, however, to prove equal to almost any demand that could be made upon it. He saw that each planet must disturb the other, and in that way he was able to render a satisfactory account of certain phenomena which had perplexed all preceding investigators. That mysterious movement by which the pole of the earth sways about among the stars had been long an unsolved enigma, but Newton showed that the moon grasped with its attraction the protuberant mass at the equatorial regions of the earth, and thus tilted the earth's axis in a way that accounted for the phenomenon which had been known but had never been explained for two thousand years. All these discoveries were brought together in that immortal work, Newton's “Principia.”

Down to the year 1687, when the “Principia “was published, Newton had lived the life of a recluse at Cambridge, being entirely occupied with those transcendent researches to which we have referred. But in that year he issued from his seclusion under circumstances of considerable historical interest. King

James the Second attempted an invasion of the rights and privileges of the University of Cambridge by issuing a command that Father Francis, a Benedictine monk, should be received as a Master of Arts in the University, without having taken the oaths of allegiance and supremacy. With this arbitrary command the University sternly refused to comply. The Vice-Chancellor was accordingly summoned to answer for an act of contempt to the authority of the Crown. Newton was one of nine delegates who were chosen to defend the independence of the University before the High Court. They were able to show that Charles the Second, who had issued a *mandamus* under somewhat similar circumstances, had been induced after due consideration to withdraw it. This argument appeared satisfactory, and the University gained their case. Newton's next step in public life was his election, by a narrow majority, as member for the University, and during the years 1688 and 1689, he seems to have attended to his parliamentary duties with considerable regularity.

An incident which happened in 1692 was apparently the cause of considerable disturbance in Newton's equanimity, if not in his health. He had gone to early morning chapel, leaving a lighted candle among his papers on his desk. Tradition asserts that his little dog "Diamond" upset the candle; at all events, when Newton came back he found that many valuable papers had perished in a conflagration. The loss of these manuscripts seems to have had a serious effect. Indeed, it has been asserted that the distress reduced Newton to a state of mental aberration for a considerable time. This has, apparently, not been confirmed, but there is no doubt that he experienced considerable disquiet, for in writing on September 13th, 1693, to Mr. Pepys, he says:

"I am extremely troubled at the embroilment I am in, and have neither ate nor slept well this twelvemonth, nor have my former consistency of mind."

Notwithstanding the fame which Newton had achieved by the publication of his "Principia," and by all his researches, the State had not as yet taken any notice whatever of the most illustrious man of science that this or any other country has ever produced. Many of his friends had exerted themselves to procure him some permanent appointment, but without success. It happened, however, that Mr. Montagu, who had sat with Newton in Parliament, was appointed Chancellor of the Exchequer in 1694. Ambitious of distinction in his new office, Mr. Montagu addressed himself to the improvement of the current coin, which was then in a very debased condition. It fortunately happened that an opportunity occurred of appointing a new official in the Mint; and Mr. Montagu on the 19th of March, 1695, wrote to offer Mr. Newton the position of warden. The salary was to be five or six hundred a year, and the business would not require more attendance than Newton could spare. The Lucasian professor accepted this post, and forthwith entered upon his new duties.

The knowledge of physics which Newton had acquired by his experiments was of much use in connection with his duties at the Mint. He carried out the re-coinage with great skill in the course of two years, and as a reward for his exertions, he was appointed, in 1697, to the Mastership of the Mint, with a salary between £1,200 and £1,500 per annum. In 1701 his duties at the Mint being so engrossing, he resigned his Lucasian professorship at Cambridge, and at the same time he had to surrender his fellowship at Trinity College. This closed his connection with the University of Cambridge. It should, however, be remarked that at a somewhat earlier stage in his career he was very nearly being appointed to an office which might have enabled the University to retain the great philosopher within its precincts. Some of his friends had almost succeeded in securing his nomination to the Provostship of King's College, Cambridge; the appointment, however, fell through, inasmuch as the statute could not be evaded, which required that the Provost of King's College should be in holy orders.

In those days it was often the custom for illustrious mathematicians, when they had discovered a solution for some new and striking problem, to publish that problem as a challenge to the world, while withholding their own solution. A famous instance of this is found in what is known as the Brachistochrone problem, which was solved by John Bernouilli. The nature of this problem may be mentioned. It was to find the shape of the curve along which a body would slide down from one point (A) to another point (B) in the shortest time. It might at first be thought that the straight line from A to B, as it is undoubtedly the shortest distance between the points, would also be the path of quickest descent; but this is not so. There is a curved line, down which a bead, let us say, would run on a smooth wire from A to B in a shorter time than the same bead would require to run down the straight wire. Bernouilli's problem was to find out what that curve must be. Newton solved it correctly; he showed that the curve was a part of what is termed a cycloid—that is to say, a curve like that which is described by a point on the rim of a carriage-wheel as the wheel runs along the ground. Such was Newton's geometrical insight that he was able to transmit a solution of the problem on the day after he had received it, to the President of the Royal Society.

In 1703 Newton, whose world-wide fame was now established, was elected President of the Royal Society. Year after year he was re-elected to this distinguished position, and his tenure, which lasted twenty-five years, only terminated with his life. It was in discharge of his duties as President of the Royal Society that Newton was brought into contact with Prince George of Denmark. In April, 1705, the Queen paid a visit to Cambridge as the guest of Dr. Bentley, the then Master of Trinity, and in a court held at Trinity Lodge on April 15th, 1705, the honour of knighthood was conferred upon the dis-coverer of gravitation.

Urged by illustrious friends, who sought the promotion of knowledge, Newton gave his attention to the publication of a new edition of the "Principia. "His duties at the Mint, however, added to the supreme duty of carrying on his original investigations, left him but little time for the more ordinary task of the revision. He was accordingly induced to associate with himself for this purpose a distinguished young mathematician, Roger Coates, a Fellow of Trinity College, Cambridge, who had recently been appointed Plumian Professor of Astronomy. On July 27th,

1713, Newton, by this time a favourite at Court, waited on the Queen, and presented her with a copy of the new edition of the "Principia."

Throughout his life Newton appears to have been greatly interested in theological studies, and he specially devoted his attention to the subject of prophecy. He left behind him a manuscript on the prophecies of Daniel and the Apocalypse of St. John, and he also wrote various theological papers. Many other subjects had from time to time engaged his attention. He studied the laws of heat; he experimented in pursuit of the dreams of the Alchymist; while the philosopher who had revealed the mechanism of the heavens found occasional relaxation in trying to interpret hieroglyphics. In the last few years of his life he bore with fortitude a painful ailment, and on Monday, March 20th, 1727, he died in the eighty-fifth year of his age.

On Tuesday, March 28th, he was buried in Westminster Abbey.

Though Newton lived long enough to receive the honour that his astonishing discoveries so justly merited, and though for many years of his life his renown was much greater than that of any of his contemporaries, yet it is not too much to say that, in the years which have since elapsed, Newton's fame has been ever steadily advancing, so that it never stood higher than it does at this moment.

We hardly know whether to admire more the sublime discoveries at which he arrived, or the extraordinary character of the intellectual processes by which those discoveries were reached. Viewed from either standpoint, Newton's "Principia" is incomparably the greatest work on science that has ever yet been produced.

- ROBERT STAWELL BALL

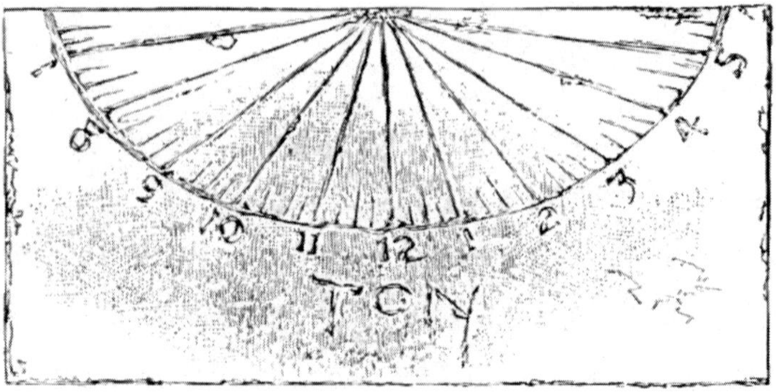

Sir Isaac Newton's Sun-dial in the Royal Society.

COMPLIMENTARY MATERIAL

A BRIEF BIOGRAPHICAL SKETCH

ISSAC NEWTON

- ROBERT STAWELL BALL

Isaac Newton

(December 25, 1642 – March 20, 1726/27)

COMPLIMENTARY MATERIAL

A BRIEF BIOGRAPHICAL SKETCH

ISSAC NEWTON

Sir Isaac Newton FRS (December 25, 1642 – March 20, 1726/27) was an English physicist and mathematician (described in his own day as a "natural philosopher") who is widely recognised as one of the most influential scientists of all time and a key figure in the scientific revolution. His book Philosophiæ Naturalis Principia Mathematica ("Mathematical Principles of Natural Philosophy"), first published in 1687, laid the foundations for classical mechanics. Newton made seminal contributions to optics, and he shares credit with Gottfried Wilhelm Leibniz for the development of calculus.

Newton's Principia formulated the laws of motion and universal gravitation, which dominated scientists' view of the physical universe for the next three centuries. By deriving Kepler's laws of planetary motion from his mathematical description of gravity, and then using the same principles to account for the trajectories of comets, the tides, the precession of the equinoxes, and other phenomena, Newton removed the last doubts about the validity of the heliocentric model of the Solar System. This work also demonstrated that the motion of objects on Earth and of celestial bodies could be described by the same principles. His prediction that Earth should be shaped as an oblate spheroid was later vindicated by the measurements of Maupertuis, La Condamine, and others, which helped convince most Continental European scientists of the superiority of Newtonian mechanics over the earlier system of Descartes.

Newton built the first practical reflecting telescope and developed a theory of colour based on the observation that a prism decomposes white light into the many colours of the visible spectrum. He formulated an empirical law of cooling, studied the speed of sound, and introduced the notion of a Newtonian fluid. In addition to his work on calculus, as a mathematician Newton contributed to the study of power series, generalised the binomial theorem to non-integer exponents, developed a method for approximating the roots of a function, and classified most of the cubic plane curves.

Newton was a fellow of Trinity College and the second Lucasian Professor of Mathematics at the University of Cambridge. He was a devout but unorthodox Christian, and, unusually for a member of the Cambridge faculty of the day, he refused to take holy orders in the Church of England, perhaps because he privately rejected the doctrine of the Trinity. Beyond his work on the mathematical sciences, Newton dedicated much of his time to the study of biblical chronology and alchemy, but most of his work in those areas remained unpublished until long after his death. In his later life, Newton became president of the Royal Society. Newton served the British government as Warden and Master of the Royal Mint.

Early Life

Isaac Newton was born according to the Julian calendar (in use in England at the time) on Christmas Day, 25 December 1642 (NS 4 January 1643), at Woolsthorpe Manor in Woolsthorpe-by-Colsterworth, a hamlet in the county of Lincolnshire. His father, also named Isaac Newton, had died three months before. Born prematurely, he was a small child; his mother Hannah Ayscough reportedly said that he could have fit inside a quart mug.

When Newton was three, his mother remarried and went to live with her new husband, the Reverend Barnabas Smith, leaving her son in the care of his maternal grandmother, Margery Ayscough. The young Isaac disliked his stepfather and maintained some enmity towards his mother for marrying him, as revealed by this entry in a list of sins committed up to the age of 19: "Threatening my father and mother Smith to burn them and the house over them." Newton's mother had three children from her second marriage.

From the age of about twelve until he was seventeen, Newton was educated at The King's School, Grantham which taught Latin and Greek but no mathematics worth mentioning. He was removed from school, and by October 1659, he was to be found at Woolsthorpe-by-Colsterworth, where his mother, widowed for a second time, attempted to make a farmer of him. Newton hated farming. Henry Stokes, master at the King's School, persuaded his mother to send him back to school so that he might complete his education. Motivated partly by a desire for revenge against a schoolyard bully, he became the top-ranked student, distinguishing himself mainly by building sundials and models of windmills. In June 1661, he was admitted to Trinity College, Cambridge, on the recommendation of his uncle Rev William Ayscough. He started as a subsizar—paying his way by performing valet's duties—until he was awarded a scholarship in 1664, which guaranteed him four more years until he would get his M.A. At that time, the college's teachings were based on those of Aristotle, whom Newton supplemented with modern philosophers such as Descartes, and astronomers such as Galileo and Thomas Street, through whom he learned of Kepler's work.

He set down in his notebook a series of 'Quaestiones' about mechanical philosophy as he found it. In 1665, he discovered the generalised binomial theorem and began to develop a mathematical theory that later became calculus. Soon after Newton had obtained his B.A. degree in August 1665, the university temporarily closed as a precaution against the Great Plague. Although he had been undistinguished as a Cambridge student, Newton's private studies at his home in Woolsthorpe over the subsequent two years saw the development of his theories on calculus, optics, and the law of gravitation.

In April 1667, he returned to Cambridge and in October was elected as a fellow of Trinity. Fellows were required to become ordained priests, although this was not enforced in the restoration years and an assertion of conformity to the Church of England was sufficient. However, by 1675 the issue could not be avoided and by then his unconventional views stood in the way. Nevertheless, Newton managed to avoid it by means of a special permission from Charles II (see "Middle years" section below).

His studies had impressed the Lucasian professor, Isaac Barrow, who was more anxious to develop his own religious and administrative potential (he became master of Trinity two years later), and in 1669, Newton succeeded him, only one year after he received his M.A. He was elected a Fellow of the Royal Society (FRS) in 1672.

Middle Years

Portrait of Newton in 1689 by Godfrey Kneller

Portrait of Newton in 1702 by Godfrey Kneller

Mathematics

Newton's work has been said "to distinctly advance every branch of mathematics then studied". His work on the subject usually referred to as fluxions or calculus, seen in a manuscript of October 1666, is now published among Newton's mathematical papers. The author of the manuscript *De analysi per aequationes numero terminorum infinitas,* sent by Isaac Barrow to John Collins in June 1669, was identified by Barrow in a letter sent to Collins in August of that year as:

Mr Newton, a fellow of our College, and very young.... but of an extraordinary genius and proficiency in these things.

Newton later became involved in a dispute with Leibniz over priority in the development of calculus (the Leibniz–Newton calculus controversy). Most modern historians believe that Newton and Leibniz developed calculus independently, although with very different notations. Occasionally it has been suggested that Newton published almost nothing about it until 1693, and did not give a full account until 1704, while Leibniz began publishing a full account of his methods in 1684. (Leibniz's notation and "differential Method", nowadays recognised as much more convenient notations, were adopted by continental European mathematicians, and after 1820 or so, also by British mathematicians.) Such a suggestion, however, fails to notice the content of calculus which critics of Newton's time and modern times have pointed out in Book 1 of Newton's Principia itself (published 1687) and in its forerunner manuscripts, such as *De motu corporum in gyrum* ("On the motion of bodies in orbit"), of 1684. The Principia is not written in the language of calculus either as we know it or as Newton's (later) 'dot' notation would write it. His work extensively uses calculus in geometric form based on limiting values of the ratios of vanishing small quantities: in the Principia itself, Newton gave demonstration of this under the name of 'the method of first and last ratios' and explained why he put his expositions in this form, remarking also that 'hereby the same thing is performed as by the method of indivisibles'.

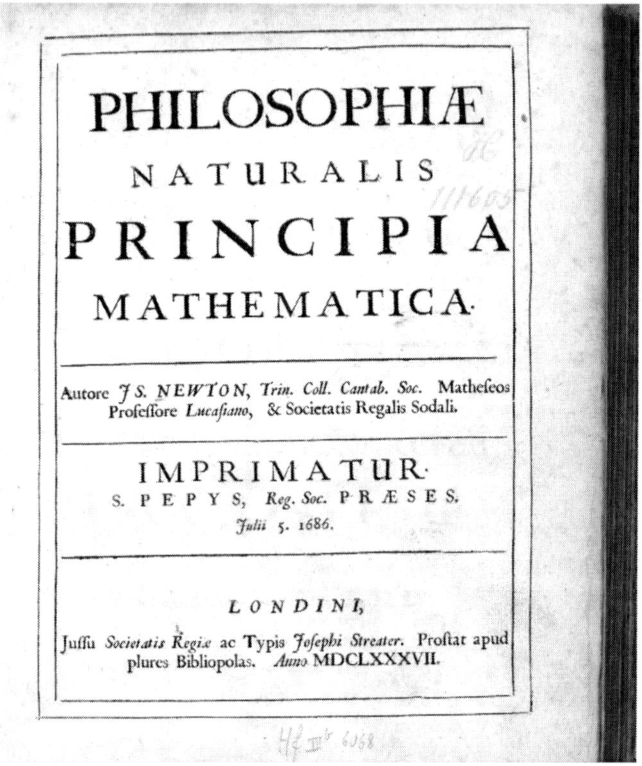

PHILOSOPHIÆ
NATURALIS
PRINCIPIA
MATHEMATICA.

Autore *JS. NEWTON*, *Trin. Coll. Cantab. Soc.* Matheseos Professore *Lucasiano*, & Societatis Regalis Sodali.

IMPRIMATUR.
S. PEPYS, *Reg. Soc.* PRÆSES.
Julii 5. 1686.

LONDINI,

Jussu *Societatis Regiæ* ac Typis *Josephi Streater*. Prostat apud plures Bibliopolas. *Anno* MDCLXXXVII.

Because of this, the Principia has been called "a book dense with the theory and application of the infinitesimal calculus" in modern times and "lequel est presque tout de ce calcul" ('nearly all of it is of this calculus') in Newton's time. His use of methods involving "one or more orders of the infinitesimally small" is present in his De motu corporum in gyrum of 1684 and in his papers on motion "during the two decades preceding 1684".

Newton had been reluctant to publish his calculus because he feared controversy and criticism. He was close to the Swiss mathematician Nicolas Fatio de Duillier. In 1691, Duillier started to write a new version of Newton's Principia, and corresponded with Leibniz. In 1693, the relationship between Duillier and Newton deteriorated and the book was never completed.

Starting in 1699, other members of the Royal Society (of which Newton was a member) accused Leibniz of plagiarism. The dispute then broke out in full force in 1711 when the Royal Society proclaimed in a study that it was Newton who was the true discoverer and labelled Leibniz a fraud. This study was cast into doubt when it was later found that Newton himself wrote the study's concluding remarks on

Leibniz. Thus began the bitter controversy which marred the lives of both Newton and Leibniz until the latter's death in 1716.

Newton is generally credited with the generalised binomial theorem, valid for any exponent. He discovered Newton's identities, Newton's method, classified cubic plane curves (polynomials of degree three in two variables), made substantial contributions to the theory of finite differences, and was the first to use fractional indices and to employ coordinate geometry to derive solutions to Diophantine equations. He approximated partial sums of the harmonic series by logarithms (a precursor to Euler's summation formula) and was the first to use power series with confidence and to revert power series. Newton's work on infinite series was inspired by Simon Stevin's decimals. A very useful modern account of Newton's mathematics was written by the foremost scholar on Newton's mathematics, D.T. Whiteside or Tom Whiteside. Tom Whiteside translated and edited all of Newton's mathematical writings and at the end of his life wrote a summing up of Newton's work and its impact. This was published in 2013 as a chapter in a book edited by Bechler.

When Newton received his MA and became a Fellow of the "College of the Holy and Undivided Trinity" in 1667, he made the commitment that "I will either set Theology as the object of my studies and will take holy orders when the time prescribed by these statutes [7 years] arrives, or I will resign from the college."

Up till this point he had not thought much about religion and had twice signed his agreement to the thirty-nine articles, the basis of Church of England doctrine.

He was appointed Lucasian Professor of Mathematics in 1669 on Barrow's recommendation. During that time, any Fellow of a college at Cambridge or Oxford was required to take holy orders and become an ordained Anglican priest. However, the terms of the Lucasian professorship required that the holder not be active in the church (presumably so as to have more time for science). Newton argued that this should exempt him from the ordination requirement, and Charles II, whose permission was needed, accepted this argument. Thus a conflict between Newton's religious views and Anglican orthodoxy was averted.

In 1666, Newton observed that the spectrum of colours exiting a prism in the position of minimum deviation is oblong, even when the light ray entering the prism is circular, which is to say, the prism refracts different colours by different angles. This led him to conclude that colour is a property intrinsic to light—a point which had been debated in prior years.

From 1670 to 1672, Newton lectured on optics. During this period he investigated the refraction of light, demonstrating that the multicoloured spectrum produced by a prism could be recomposed into white light by a lens and a second prism. Modern scholarship has revealed that Newton's analysis and resynthesis of white light owes a debt to corpuscular alchemy.

He also showed that coloured light does not change its properties by separating out a coloured beam and shining it on various objects. Newton noted that regardless of whether it was reflected, scattered, or transmitted, it remained the same colour. Thus, he observed that colour is the result of objects interacting with already-coloured light rather than objects generating the colour themselves. This is known as Newton's theory of colour.

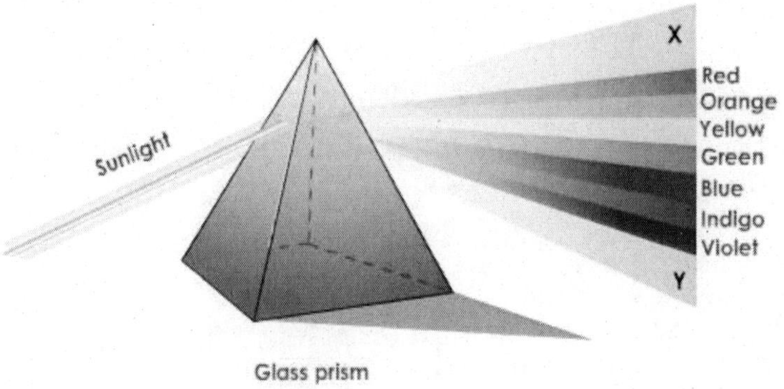

Illustration of a dispersive prism decomposing white light into the colours of the spectrum, as discovered by Newton.

From this work, he concluded that the lens of any refracting telescope would suffer from the dispersion of light into colours (chromatic aberration). As a proof of the concept, he constructed a telescope using reflective mirrors instead of lenses as the objective to bypass that problem. Building the design, the first known functional reflecting telescope, today known as a Newtonian telescope, involved solving the problem of a suitable mirror material and shaping technique. Newton ground his own mirrors out of a custom composition of highly reflective speculum metal, using Newton's rings to judge the quality of the optics for his telescopes. In late 1668 he was able to produce this first reflecting telescope. It was about eight inches long and it gave a clearer and larger image.

In 1671, the Royal Society asked for a demonstration of his reflecting telescope. Their interest encouraged him to publish his notes, Of Colours, which he later expanded into the work Opticks. When Robert Hooke criticised some of Newton's ideas, Newton was so offended that he withdrew from public debate.

Replica of Newton's second Reflecting telescope that he presented to the Royal Society in 1672.

Newton and Hooke had brief exchanges in 1679–80, when Hooke, appointed to manage the Royal Society's correspondence, opened up a correspondence intended to elicit contributions from Newton to Royal Society transactions, which had the effect of stimulating Newton to work out a proof that the elliptical form of planetary orbits would result from a centripetal force

- ROBERT STAWELL BALL

inversely proportional to the square of the radius vector (see Newton's law of universal gravitation – History and De motu corporum in gyrum). But the two men remained generally on poor terms until Hooke's death.

Newton argued that light is composed of particles or corpuscles, which were refracted by accelerating into a denser medium. He verged on soundlike waves to explain the repeated pattern of reflection and transmission by thin films (Opticks Bk.II, Props. 12), but still retained his theory of 'fits' that disposed corpuscles to be reflected or transmitted (Props.13). However, later physicists favoured a purely wavelike explanation of light to account for the interference patterns and the general phenomenon of diffraction. Today's quantum mechanics, photons, and the idea of wave–particle duality bear only a minor resemblance to Newton's understanding of light.

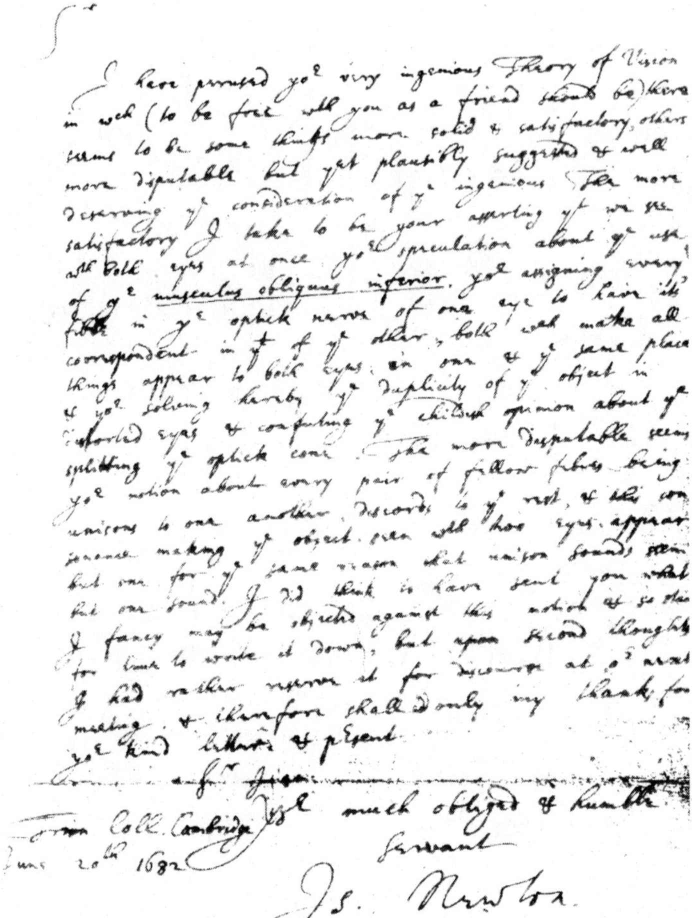

Sr

I have perused yor very ingenious Theory of Vision
in wch (to be free wth you as a friend should be) there
seems to be some things more solid & satisfactory, others
more disputable but yet plausibly suggested & well
deserving ye consideration of ye ingenious. The more
satisfactory I take to be your asserting yt we see
wth both eyes at once, yor speculation about ye use
of ye musculus obliquus inferior, yor assigning every
fibre in ye optick nerve of one eye to have its
correspondent in yt of ye other, both wch make all
things appear to both eyes in one & ye same place
& yor solving hereby ye duplicity of ye object in
distorted eyes & confuting ye childish opinion about ye
splitting ye optick cone. The more disputable seems
yor notion about every pair of fellow fibres being
unisons to one another, discords to ye rest, & this con-
sonance making ye object seen wth two eyes appear
but one for ye same reason that unison sounds seem
but one sound. I did think to have sent you what
I fancy may be objected against this notion & so the
time to write it down, but upon second thoughts
I had rather reserve it for discourse at or next
meeting, & therefore shall add only my thanks for
yor kind letters & present.

~~Sr~~

Trin. Coll. Cambridge
June 20th 1682

Yor much obliged & humble
Servant
Is. Newton.

Facsimile of a 1682 letter from Isaac Newton
to Dr. William Briggs, commenting on Briggs'
"A New Theory of Vision"

In his Hypothesis of Light of 1675, Newton posited the existence of the ether to transmit forces between particles. The contact with the theosophist Henry More, revived his interest in alchemy. He replaced the ether with occult forces based on Hermetic ideas of attraction and repulsion between particles. John Maynard Keynes, who acquired many of Newton's writings on alchemy, stated that "Newton was not the first of the age of reason: He was the last of the magicians." Newton's interest in alchemy cannot be isolated from his contributions to science. This was at a time when there was no clear distinction between alchemy and science. Had he not relied on the occult idea of action at a distance, across a vacuum, he might not have developed his theory of gravity. (See also Isaac Newton's occult studies.)

In 1704, Newton published Opticks, in which he expounded his corpuscular theory of light. He considered light to be made up of extremely subtle corpuscles, that ordinary matter was made of grosser corpuscles and speculated that through a kind of alchemical transmutation "Are not gross Bodies and Light convertible into one another, ... and may not Bodies receive much of their Activity from the Particles of Light which enter their Composition?" Newton also constructed a primitive form of a frictional electrostatic generator, using a glass globe.

In an article entitled "Newton, prisms, and the 'opticks' of tunable lasers" it is indicated that Newton in his book Opticks was the first to show a diagram using a prism as a beam expander. In the same book he describes, via diagrams, the use of multiple-prism arrays. Some 278 years after Newton's discussion, multiple-prism beam expanders became central to the development of narrow-linewidth tunable lasers. Also, the use of these prismatic beam expanders led to the multiple-prism dispersion theory.

Subsequent to Newton, much has been amended. Young and Fresnel combined Newton's particle theory with Huygens' wave theory to show that colour is the visible manifestation of light's wavelength.

Science also slowly came to realise the difference between perception of colour and mathematisable optics. The German poet and scientist, Goethe, could not shake the Newtonian foundation but "one hole Goethe did find in Newton's armour, ... Newton had committed himself to the doctrine that refraction without colour was impossible. He therefore thought that the object-glasses of telescopes must forever remain imperfect, achromatism and refraction being incompatible. This inference was proved by Dollond to be wrong."

Mechanics and Gravitation

In 1679, Newton returned to his work on (celestial) mechanics by considering gravitation and its effect on the orbits of planets with reference to Kepler's laws of planetary motion. This followed stimulation by a brief exchange of letters in 1679–80 with Hooke, who had been appointed to manage the Royal Society's correspondence, and who opened a correspondence intended to elicit contributions from Newton to Royal Society transactions. Newton's reawakening interest in astronomical matters received further stimulus by the appearance of a comet in the winter of 1680–1681, on which he corresponded with John Flamsteed. After the exchanges with Hooke, Newton worked out proof that the elliptical form of planetary orbits would result from a centripetal force inversely proportional to the square of the radius vector (see Newton's law of universal gravitation – History and *De motu corporum in gyrum*).

- ROBERT STAWELL BALL

Newton communicated his results to Edmond Halley and to the Royal Society in *De motu corporum in gyrum*, a tract written on about nine sheets which was copied into the Royal Society's Register Book in December 1684. This tract contained the nucleus that Newton developed and expanded to form the Principia.

The Principia was published on 5 July 1687 with encouragement and financial help from Edmond Halley. In this work, Newton stated the three universal laws of motion. Together, these laws describe the relationship between any object, the forces acting upon it and the resulting motion, laying the foundation for classical mechanics. They contributed to many advances during the Industrial Revolution which soon followed and were not improved upon for more than 200 years. Many of these advancements continue to be the underpinnings of non-relativistic technologies in the modern world.

He used the Latin word gravitas (weight) for the effect that would become known as gravity, and defined the law of universal gravitation.

In the same work, Newton presented a calculus-like method of geometrical analysis using 'first and last ratios', gave the first analytical determination (based on Boyle's law) of the speed of sound in air, inferred the oblateness of Earth's spheroidal figure, accounted for the precession of the equinoxes as a result of the Moon's gravitational attraction on the Earth's oblateness, initiated the gravitational study of the irregularities in the motion of the moon, provided a theory for the determination of the orbits of comets, and much more.

Newton made clear his heliocentric view of the Solar System—developed in a somewhat modern way, because already in the mid-1680s he recognised the "deviation of the Sun" from the centre of gravity of the Solar System. For Newton, it was not precisely the centre of the Sun or any other body that could be considered at rest, but rather "the common centre of gravity of the Earth, the Sun and all the Planets is to be esteem'd the Centre of the World", and this centre of gravity "either is at rest or moves uniformly forward in a right line" (Newton adopted the "at rest" alternative in view of common consent that the centre, wherever it was, was at rest).

Newton's postulate of an invisible force able to act over vast distances led to him being criticised for introducing "occult agencies" into science. Later, in the second edition of the Principia (1713), Newton firmly rejected such criticisms in a concluding General Scholium, writing that it was enough that the phenomena implied a gravitational attraction, as they did; but they did not so far indicate its cause, and it was both unnecessary and improper to frame hypotheses of things that were not implied by the phenomena. (Here Newton used what became his famous expression "hypotheses non-fingo").

With the Principia, Newton became internationally recognised. He acquired a circle of admirers, including the Swiss-born mathematician Nicolas Fatio de Duillier.

Classification of Cubics

Descartes was the most important early influence on Newton the mathematician. Descartes freed plane curves from the Greek and Macedonian limitation to conic sections, and Newton followed his lead by classifying the cubic curves in the plane. He found 72 of the 78 species of cubics. He also divided them into four types, satisfying different equations, and in 1717 Stirling, probably with Newton's help, proved that every cubic was one of these four types. Newton also claimed that the four types could be obtained by plane projection from one of them, and this was proved in 1731.

- ROBERT STAWELL BALL

Later Life

Isaac Newton in old age in 1712,
portrait by Sir James Thornhill.

In the 1690s, Newton wrote a number of religious tracts dealing with the literal and symbolic interpretation of the Bible. A manuscript Newton sent to John Locke in which he disputed the fidelity of 1 John 5:7 and its fidelity to the original manuscripts of the New Testament, remained unpublished until 1785.

Even though a number of authors have claimed that the work might have been an indication that Newton disputed the belief in Trinity, others assure that Newton did question the passage but never denied Trinity as such. His biographer, scientist Sir David Brewster, who compiled his manuscripts for over 20 years, wrote about the controversy in well-known book Memoirs of the Life, Writings, and Discoveries of Sir Isaac Newton, where he explains that Newton questioned the veracity of those passages, but he never denied the doctrine of Trinity as such. Brewster states that Newton was never known as an Arian during his lifetime, it was first William Whiston (an Arian) who argued that "Sir Isaac Newton was so hearty for the Baptists, as well as for the

Eusebians or Arians, that he sometimes suspected these two were the two witnesses in the Revelations," while other like Hopton Haynes (a Mint employee and Humanitarian), "mentioned to Richard Baron, that Newton held the same doctrine as himself".

Later works—The Chronology of Ancient Kingdoms Amended (1728) and Observations Upon the Prophecies of Daniel and the Apocalypse of St. John (1733)—were published after his death. He also devoted a great deal of time to alchemy (see above).

Newton was also a member of the Parliament of England for Cambridge University in 1689–90 and 1701–2, but according to some accounts his only comments were to complain about a cold draught in the chamber and request that the window be closed. He was however noted by Cambridge diarist Abraham de la Pryme as having rebuked students who were frightening local residents by claiming that a house was haunted.

Newton moved to London to take up the post of warden of the Royal Mint in 1696, a position that he had obtained through the patronage of Charles Montagu, 1st Earl of Halifax, then Chancellor of the Exchequer. He took charge of England's great recoining, somewhat treading on the toes of Lord Lucas, Governor of the Tower (and securing the job of deputy comptroller of the temporary Chester branch for Edmond Halley). Newton became perhaps the best-known Master of the Mint upon the death of Thomas Neale in 1699, a position Newton held for the last 30 years of his life. These appointments were intended as sinecures, but Newton took them seriously, retiring from his Cambridge duties in 1701, and exercising his power to reform the currency and punish clippers and counterfeiters.

As Warden, and afterwards Master, of the Royal Mint, Newton estimated that 20 percent of the coins taken in during the Great Recoinage of 1696 were counterfeit. Counterfeiting was high treason, punishable by the felon being hanged, drawn and quartered. Despite this, convicting even the most flagrant criminals could be extremely difficult. However, Newton proved equal to the task.

Disguised as a habitué of bars and taverns, he gathered much of that evidence himself. For all the barriers placed to prosecution, and separating the branches of government, English law still had ancient and formidable customs of authority. Newton had himself made a justice of the peace in all the home counties—there is a draft of a letter regarding this matter stuck into Newton's personal first edition of his Philosophiæ Naturalis Principia Mathematica which he must have been amending at the time. Then he conducted more than 100 cross-examinations of witnesses, informers, and suspects between June 1698 and Christmas 1699. Newton successfully prosecuted 28 coiners. As a result of a report written by Newton on 21 September 1717 to the Lords Commissioners of His Majesty's Treasury the bimetallic relationship between gold coins and silver coins was changed by Royal proclamation on 22 December 1717, forbidding the exchange of gold guineas for more than 21 silver shillings. This inadvertently resulted in a silver shortage as silver coins were used to pay for imports, while exports were paid for in gold, effectively moving Britain from the silver standard to its first gold standard. It is a matter of debate as whether he intended to do this or not. It has been argued that Newton conceived of his work at the Mint as a continuation of his alchemical work.

Newton was made President of the Royal Society in 1703 and an associate of the French Académie des Sciences. In his position at the Royal Society, Newton made an enemy of John Flamsteed, the Astronomer Royal, by prematurely publishing Flamsteed's Historia Coelestis Britannica, which Newton had used in his studies.

In April 1705, Queen Anne knighted Newton during a royal visit to Trinity College, Cambridge. The knighthood is likely to have been motivated by political considerations connected with the Parliamentary election in May 1705, rather than any recognition of Newton's scientific work or services as Master of the Mint. Newton was the second scientist to be knighted, after Sir Francis Bacon.

Newton was one of many people who lost heavily when the South Sea Company collapsed. Their most significant trade was slaves, and according to his niece, he lost around £20,000.

Towards the end of his life, Newton took up residence at Cranbury Park, near Winchester with his niece and her husband, until his death in 1727. His half-niece, Catherine Barton Conduitt, served as his hostess in social affairs at his house on Jermyn Street in London; he was her "very loving Uncle", according to his letter to her when she was recovering from smallpox.

Newton died in his sleep in London on 20 March 1727 (OS 20 March 1726; NS 31 March 1727) and was buried in Westminster Abbey. Voltaire may have been present at his funeral. A bachelor, he had divested much of his estate to relatives during his last years, and died intestate. His papers went to John Conduitt and Catherine Barton. After his death, Newton's hair was examined and found to contain mercury, probably resulting from his alchemical pursuits. Mercury poisoning could explain Newton's eccentricity in late life.

- ROBERT STAWELL BALL

Newton's tomb in Westminster Abbey.

Personal Relations

Although it was claimed that he was once engaged, Newton never married. The French writer and philosopher Voltaire, who was in London at the time of Newton's funeral, said that he "was never sensible to any passion, was not subject to the common frailties of mankind, nor had any commerce with women—a circumstance which was assured me by the physician and surgeon who attended him in his last moments". The widespread belief that he died a virgin has been commented on by writers such as mathematician Charles Hutton, economist John Maynard Keynes, and physicist Carl Sagan.

Newton did have a close friendship with the Swiss mathematician Nicolas Fatio de Duillier, whom he met in London around 1690. Their intense relationship came to an abrupt and unexplained end in 1693, and at the same time Newton suffered a nervous breakdown. Some of their correspondence has survived.

In September of that year, Newton had a breakdown which included sending wild accusatory letters to his friends Samuel Pepys and John Locke. His note to the latter included the charge that Locke *"endeavoured to embroil me with woemen"*.

After Death

The mathematician Joseph-Louis Lagrange often said that Newton was the greatest genius who ever lived, and once added that Newton was also "the most fortunate, for we cannot find more than once a system of the world to establish." English poet Alexander Pope was moved by Newton's accomplishments to write the famous epitaph:

"Nature and nature's laws lay hid in night;
God said "Let Newton be" and all was light."

Newton himself had been rather more modest of his own achievements, famously writing in a letter to Robert Hooke in February 1676:

"If I have seen further it is by standing on the shoulders of giants."

Two writers think that the above quotation, written at a time when Newton and Hooke were in dispute over optical discoveries, was an oblique attack on

Hooke (said to have been short and hunchbacked), rather than—or in addition to—a statement of modesty. On the other hand, the widely known proverb about standing on the shoulders of giants, published among others by seventeenth-century poet George Herbert (a former orator of the University of Cambridge and fellow of Trinity College) in his Jacula Prudentum (1651), had as its main point that "a dwarf on a giant's shoulders sees farther of the two", and so its effect as an analogy would place Newton himself rather than Hooke as the 'dwarf'.

In a later memoir, Newton wrote:

"I do not know what I may appear to the world, but to myself I seem to have been only like a boy playing on the sea-shore, and diverting myself in now and then finding a smoother pebble or a prettier shell than ordinary, whilst the great ocean of truth lay all undiscovered before me."

In 1816, a tooth said to have belonged to Newton was sold for £730 (us$3,633) in London to an aristocrat who had it set in a ring. The Guinness World Records 2002 classified it as the most valuable tooth, which would value approximately £25,000 (us$35,700) in late 2001. Who bought it and who currently has it has not been disclosed.

Albert Einstein kept a picture of Newton on his study wall alongside ones of Michael Faraday and James Clerk Maxwell. Newton remains influential to today's scientists, as demonstrated by a 2005 survey of members of Britain's Royal Society (formerly headed by Newton) asking who had the greater effect on the history of science, Newton or Einstein. Royal Society scientists deemed Newton to have made the greater overall contribution.

In 1999, an opinion poll of 100 of today's leading physicists voted Einstein the "greatest physicist ever;" with Newton the runner-up, while a parallel survey of rank-and-file physicists by the site: **PhysicsWorld** (http://physicsworld.com) gave the top spot to Newton.

Commemorations

A statue of Isaac Newton, looking at an apple at his feet, can be seen at the Oxford University Museum of Natural History.

Newton statue on display at the Oxford University Museum of Natural History Newton's monument (1731) can be seen in Westminster Abbey, at the north of the entrance to the choir against the choir screen, near his tomb. It was executed by the sculptor Michael Rysbrack (1694–1770) in white and grey marble with design by the architect William Kent. The monument features a figure of Newton reclining on top of a sarcophagus, his right elbow resting on several of his great books and his left hand pointing to a scroll with a mathematical design. Above him is a pyramid and a celestial globe showing the signs of the Zodiac and the path of the comet of 1680. A relief panel depicts putti using instruments such as a telescope and prism. The Latin inscription on the base translates as:

Here is buried Isaac Newton, Knight, who by a strength of mind almost divine, and mathematical principles peculiarly his own, explored the course and figures of the planets, the paths of comets, the tides of the sea, the dissimilarities in rays of light, and, what no other scholar has previously imagined, the properties of the colours thus produced. Diligent, sagacious and faithful, in his expositions of nature, antiquity and the holy Scriptures, he vindicated by his philosophy the majesty of God mighty and good, and expressed the simplicity of the Gospel in his manners. Mortals rejoice that there has existed such and so great an ornament of the human race! He was born on 25 December 1642, and died on 20 March 1726/7.—Translation from G.L. Smyth, The Monuments and Genii of St. Paul's Cathedral, and of Westminster Abbey (1826).

From 1978 until 1988, an image of Newton designed by Harry Ecclestone appeared on Series D £1 banknotes issued by the Bank of England (the last £1 notes to be issued by the Bank of England). Newton was shown on the reverse of the notes holding a book and accompanied by a telescope, a prism and a map of the Solar System. Eduardo Paolozzi's Newton, after William Blake (1995), outside the British Library in London.

Religious Views

Although born into an Anglican family, by his thirties Newton held a Christian faith that, had it been made public, would not have been considered orthodox by mainstream Christianity; in recent times he has been described as a heretic.

By 1672 he had started to record his theological researches in notebooks which he showed to no one and which have only recently been examined. They demonstrate an extensive knowledge of early church writings and show that in the conflict between Athanasius and Arius which defined the Creed, he took the side of Arius, the loser, who rejected the conventional view of the Trinity. Newton "recognized Christ as a divine mediator between God and man, who was subordinate to the Father who created him." He was especially interested in prophecy, but for him, "the great apostasy was trinitarianism."

- ROBERT STAWELL BALL

Newton tried unsuccessfully to obtain one of the two fellowships that exempted the holder from the ordination requirement. At the last moment in 1675 he received a dispensation from the government that excused him and all future holders of the Lucasian chair.

In Newton's eyes, worshipping Christ as God was idolatry, to him the fundamental sin. Historian Stephen D. Snobelen says of Newton, "Isaac Newton was a heretic. But he never made a public declaration of his private faith—which the orthodox would have deemed extremely radical. He hid his faith so well that scholars are still unravelling his personal beliefs." Snobelen concludes that Newton was at least a Socinian sympathiser (he owned and had thoroughly read at least eight Socinian books), possibly an Arian and almost certainly an anti-trinitarian.

In a minority view, T.C. Pfizenmaier argues that Newton held the Eastern Orthodox view on the Trinity. However, this type of view 'has lost support of late with the availability of Newton's theological papers', and now most scholars identify Newton as an Antitrinitarian monotheist.

Although the laws of motion and universal gravitation became Newton's best-known discoveries, he warned against using them to view the Universe as a mere machine, as if akin to a great clock. He said, "Gravity explains the motions of the planets, but it cannot explain who set the planets in motion. God governs all things and knows all that is or can be done."

Along with his scientific fame, Newton's studies of the Bible and of the early Church Fathers were also noteworthy. Newton wrote works on textual criticism, most notably An Historical Account of Two Notable Corruptions of Scripture. He placed the crucifixion of Jesus Christ at 3 April, AD 33, which agrees with one traditionally accepted date.

He believed in a rationally immanent world, but he rejected the hylozoism implicit in Leibniz and Baruch Spinoza. The ordered and dynamically informed

Universe could be understood, and must be understood, by an active reason. In his correspondence, Newton claimed that in writing the Principia "I had an eye upon such Principles as might work with considering men for the belief of a Deity". He saw evidence of design in the system of the world: "Such a wonderful uniformity in the planetary system must be allowed the effect of choice". But Newton insisted that divine intervention would eventually be required to reform the system, due to the slow growth of instabilities. For this, Leibniz lampooned him: "God Almighty wants to wind up his watch from time to time: otherwise it would cease to move. He had not, it seems, sufficient foresight to make it a perpetual motion."

Newton's position was vigorously defended by his follower Samuel Clarke in a famous correspondence. A century later, Pierre-Simon Laplace's work "Celestial Mechanics" had a natural explanation for why the planet orbits don't require periodic divine intervention.

Effect on Religious Thought

Newton and Robert Boyle's approach to the mechanical philosophy was promoted by rationalist pamphleteers as a viable alternative to the pantheists and enthusiasts, and was accepted hesitantly by orthodox preachers as well as dissident preachers like the latitudinarians. The clarity and simplicity of science was seen as a way to combat the emotional and metaphysical superlatives of both superstitious enthusiasm and the threat of atheism, and at the same time, the second wave of English deists used Newton's discoveries to demonstrate the possibility of a "Natural Religion".

Newton, by William Blake; here, Newton is depicted critically as a "divine geometer". This copy of the work is currently held by the Tate Collection.

The attacks made against pre-Enlightenment "magical thinking", and the mystical elements of Christianity, were given their foundation with Boyle's mechanical conception of the Universe. Newton gave Boyle's ideas their completion through mathematical proofs and, perhaps more importantly, was very successful in popularising them.

Occult

In a manuscript he wrote in 1704 in which he describes his attempts to extract scientific information from the Bible, he estimated that the world would end no earlier than 2060. In predicting this he said, "This I mention not to assert when the time of the end shall be, but to put a stop to the rash conjectures of fanciful men who are frequently predicting the time of the end, and by doing so bring the sacred prophesies into discredit as often as their predictions fail."

Alchemy

In the character of Morton Opperly in "Poor Superman" (1951), speculative fiction author Fritz Leiber says of Newton, "Everyone knows Newton as the great scientist. Few remember that he spent half his life muddling with alchemy, looking for the philosopher's stone. That was the pebble by the seashore he really wanted to find."

Of an estimated ten million words of writing in Newton's papers, about one million deal with alchemy. Many of Newton's writings on alchemy are copies of other manuscripts, with his own annotations. Alchemical texts mix artisanal knowledge with philosophical speculation, often hidden behind layers of wordplay, allegory, and imagery to protect craft secrets. Some of the content contained in Newton's papers could have been considered heretical by the church.

In 1888, after spending sixteen years cataloging Newton's papers, Cambridge University kept a small number and returned the rest to the Earl of Portsmouth. In 1936, a descendant offered the papers for sale at Sotheby's. The collection was broken up and sold for a total of about £9,000. John Maynard Keynes was one of about three dozen bidders who obtained part of the collection at auction. Keynes went on to reassemble an estimated half of Newton's collection of papers on alchemy before donating his collection to Cambridge University in 1946.

Enlightenment Philosophers

Enlightenment philosophers chose a short history of scientific predecessors – Galileo, Boyle, and Newton principally – as the guides and guarantors of their applications of the singular concept of Nature and Natural law to every physical and social field of the day. In this respect, the lessons of history and the social structures built upon it could be discarded. It was Newton's conception of the Universe based upon Natural and rationally understandable laws that became one of the seeds for Enlightenment ideology. Locke and Voltaire applied concepts of Natural Law to political systems advocating intrinsic rights; the physiocrats and Adam Smith applied Natural conceptions of psychology and self-interest to economic systems; and sociologists criticised the current social order for trying to fit history into Natural models of progress. Monboddo and Samuel Clarke resisted elements of Newton's work, but eventually rationalised it to conform with their strong religious views of nature.

- ROBERT STAWELL BALL

Apple Incident

Reputed descendants of Newton's apple tree, (from top to bottom) at Trinity College, Cambridge, the Cambridge University Botanic Garden, and the Instituto Balseiro library garden

Newton himself often told the story that he was inspired to formulate his theory of gravitation by watching the fall of an apple from a tree. Although it has been said that the apple story is a myth and that he did not arrive at his theory of gravity in any single moment, acquaintances of Newton (such as William Stukeley, whose manuscript account of 1752 has been made available by the Royal Society) do in fact confirm the incident, though not the cartoon version that the apple actually hit Newton's head. Stukeley recorded in his Memoirs of Sir Isaac Newton's Life a conversation with Newton in Kensington on 15 April 1726:

"We went into the garden, & drank thea under the shade of some appletrees; only

he, & my self. amidst other discourse, he told me, he was just in the same situation, as when formerly, the notion of gravitation came into his mind. "why should that apple always descend perpendicularly to the ground," thought he to himself; occasion'd by the fall of an apple, as he sat in a contemplative mood. "why should it not go sideways, or upwards? but constantly to the earths center? assuredly, the reason is, that the earth draws it. there must be a drawing power in matter. & the sum of the drawing power in the matter of the earth must be in the earths center, not in any side of the earth. therefore dos this apple fall perpendicularly, or toward the center. if matter thus draws matter; it must be in proportion of its quantity. therefore the apple draws the earth, as well as the earth draws the apple."

John Conduitt, Newton's assistant at the Royal Mint and husband of Newton's niece, also described the event when he wrote about Newton's life:

"In the year 1666 he retired again from Cambridge to his mother in Lincolnshire. Whilst he was pensively meandering in a garden it came into his thought that the power of gravity (which brought an apple from a tree to the ground) was not limited to a certain distance from earth, but that this power must extend much further than was usually thought. Why not as high as the Moon said he to himself & if so, that must influence her motion & perhaps retain her in her orbit, whereupon he fell a calculating what would be the effect of that supposition."

In similar terms, Voltaire wrote in his Essay on Epic Poetry (1727), "Sir Isaac Newton walking in his gardens, had the first thought of his system of gravitation, upon seeing an apple falling from a tree."

It is known from his notebooks that Newton was grappling in the late 1660s with the idea that terrestrial gravity extends, in an inverse-square proportion, to the Moon; however it took him two decades to develop the full-fledged theory. The question was not whether gravity existed, but whether it extended so far from Earth that it could also be the force holding the Moon to its orbit. Newton showed that if the force decreased as the inverse square of the distance, one could indeed calculate the Moon's orbital period, and get good agreement. He guessed the same force was responsible for other orbital motions, and hence named it "universal gravitation".

Various trees are claimed to be "the" apple tree which Newton describes. The King's School, Grantham, claims that the tree was purchased by the school, uprooted and transported to the headmaster's garden some years later. The staff of the [now] National Trust-owned Woolsthorpe Manor dispute this, and claim that a tree present in their gardens is the one described by Newton. A descendant of the original tree can be seen growing outside the main gate of Trinity College, Cambridge, below the room Newton lived in when he studied there. The National Fruit Collection at Brogdale can supply grafts from their tree, which appears identical to Flower of Kent, a coarse-fleshed cooking variety.

- ROBERT STAWELL BALL

ABOUT THE AUTHOR

ROBERT STAWELL BALL

(July 1, 1840 – November 25, 1913)

Sir Robert Stawell Ball was an Irish astronomer who founded the screw theory.

He was the son of naturalist Robert Ball and Amelia Gresley Hellicar. He was born in Dublin.

Ball worked for Lord Rosse from 1865 to 1867. In 1867 he became Professor of Applied Mathematics at the Royal College of Science in Dublin. There he lectured on mechanics and published an elementary account of the science.

In 1874 Ball was appointed Royal Astronomer of Ireland and Andrews Professor of Astronomy in the University of Dublin at Dunsink Observatory.

Ball contributed to the science of kinematics by delineating the screw displacement:

When Ball and the screw theorists speak of screws they no longer mean actual cylindrical objects with helical threads cut into them but the possible motion of any body whatsoever, including that of the screw independently of the nut.

Ball's treatise *The Theory of Screws* (1876) is now in the public domain. His work on screw dynamics earned him in 1879 the

Cunningham Medal of the Royal Irish Academy.

In 1882 Popular Science Monthly carried his article "*A Glimpse through the Corridors of Time*". The following year it carried his two-part article on "*The Boundaries of Astronomy*".

Ball expounded the tides in *Time and Tide*: a Romance of the Moon In 1892 he was appointed Lowndean Professor of Astronomy and Geometry at Cambridge University at the same time becoming director of the Cambridge Observatory. He was a fellow of King's College, Cambridge.

In 1900 Cambridge University Press published *A Treatise on the Theory of Screws*. That year he also published *The Story of the Heavens*. Much in the limelight, he stood as President of the Quaternion Society. He was also President of the Mathematical Association in 1900.

In 1908 he published *A Treatise on Spherical Astronomy*, which is a textbook on astronomy starting from spherical trigonometry and the celestial sphere, considering atmospheric refraction and aberration of light, and introducing basic use of a generalised instrument.

His work *The Story of the Heavens* is mentioned in the "Ithaka" chapter of Ulysses. His lectures, articles and books (e.g. *Starland* and *The Story of the Heavens*) were mostly popular and simple in style.

He died in Cambridge and was buried at the Parish of the Ascension Burial Ground in Cambridge, with his wife Lady Francis Elizabeth Ball. Their children were: Frances Amelia, Robert Steele, William Valentine (later Sir), Mary Agnetta, Charles Rowan Hamilton, and Randall Gresley (later Colonel).

LECTURES

Ball became celebrated for his popular lectures on science. He gave an estimated 2500 lectures between 1875 and 1910 in towns and cities across Britain and Ireland. In 1892, 1898 and 1900 he was invited to deliver the Royal Institution Christmas Lecture. *Astronomy; Astronomy* and *Great Chapters from the Book of Nature.*

FEEDBACK

Now that you have read the book ...

Was it interesting?

Did you enjoy what you wanted to read?
Was there any room for improvement?

Let us know at:

http://www.diamondbooks.ca/feedback

Your feedback is highly appreciated.

Thank you!

Made in United States
Cleveland, OH
06 November 2026

47647256R00078